First Edition

Genuine Autographed Collectible

Do you want me to sign it in ink or in lipstick?

NEW SCIENTIFIC THEORY

SPERM MANIFESTO

10 RULES FOR THE ROAD

There Is Only Room for **ONE** at The Top

Gift Card

Date:

To:

From:

Message:

What Do Books Do?

BOOKS ARE POWERFUL

Books **E**ducate!
Books **E**nlighten!
Books **E**mpower!
Books **E**mancipate!
Books **E**ntertain!
Books Spring **E**ternal!
Books Drive **E**xploration!
Books Spark **E**volution!
Books Ignite **Re**volution!

Sharon Esther Lampert

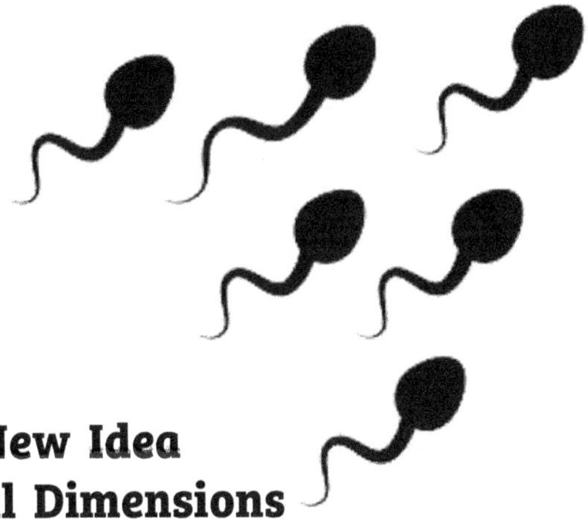

A Mind Stretched to a New Idea
Never Returns to Its Original Dimensions

— Oliver Wendell Holmes

NEW SCIENTIFIC THEORY

SPERM MANIFESTO

The Drama that Occurs Inside the Womb Is the
Same Drama that Occurs Outside the Womb
There Is Only Room for **ONE** at The Top

Prodigy Sharon Esther Lampert
Genius: The Gift of Divine Revelation
BEZALEL: EXODUS 31: 1-3

NUMERO UNO

Science, Biology, Self-Help, Philosophy, Spirituality, Religion, Philosopher Queen Sharon Esther Lampert

SPERM MANIFESTO: 10 RULES FOR THE ROAD

KADIMAH PRESS
Gifts of Genius

Books may be purchased for education, business, or sales promotional use.

ISBN Hardcover: 979-8-3305-2236-1
ISBN Paperback: 979-8-3305-2238-5
ISBN E-Book: 979-8-3305-2237-8
Library of Congress Catalog Card Number: 2024923329

FAN MAIL:
SharonEstherLampert.com
FANS@SharonEstherLampert.com

Cover and Interior Book Design: Creative Genius Sharon Esther Lampert
Editor: Dave Segal
Palm Beach Book Publisher, Phone: 917-767-5843
Sharon@PalmBeachBookPublisher.com

To Order Book:
Ingram, 1 Ingram Blvd. La Vergne, TN 37086-3629
Phone: 615-793-5000
Fax orders: 615-287-6990

First Edition

Manufactured in the United States of America

Age 9
THE QUEEN HAS ARRIVED
My daughter is a poet, philosopher, and teacher.
She is the Princess & Pea!
BEAUTY & BRAINS!
LOVE & XOXO
MOMMY

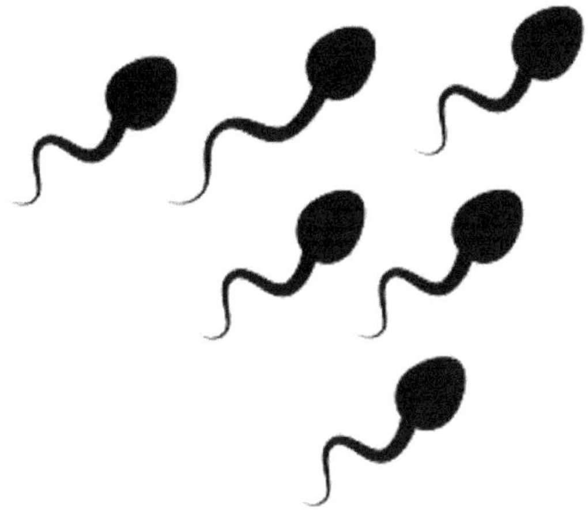

NEW SCIENTIFIC THEORY

SPERM MANIFESTO
10 RULES FOR THE ROAD
There Is Only Room for **ONE** at The Top

NUMERO UNO

KADIMAH PRESS
Gifts of Genius

BE ART

ART IS SMART
ART IS OF THE HEART
MAKE ART NOT WAR
YOU ARE BORN FOR GREATNESS
YOU ARE A MASTERPIECE

Sharon Esther Lampert

DEDICATIONS

My Muse Dr. Karl Bardosh

THE GOAT: THE GREATEST OF ALL TIME
KING BIBI: Prime Minister Benjamin Netanyahu
The Greatest World Leader Since The Dawn of History

"YOU CANNOT NEGOTIATE WITH EVIL
DESTROY THE EVIL OR BE DESTROYED BY THE EVIL"

SHARON ESTHER LAMPERT
PRINCESS KADIMAH: 8TH PROPHETESS OF ISRAEL

To My Gift The Creative Apparatus

"To a Mind
That Is Still
The Whole
Universe
Surrenders"

— Lau Tzu

"Talent Hits a Target
No One Else Can Hit
Genius Hits a Target
No One Else Can See"

— Arthur Schopenhauer

MOMMY
LOVE OF MY LIFETIME
WHO KNEW WHO I WAS
FROM THE INSIDE OUT!

"I Have Nothing
to Declare Except
My Genius"

— Oscar Wilde

NEW SCIENTIFIC THEORY

SPERM MANIFESTO

THERE IS ONLY ROOM FOR **ONE** AT THE TOP

NUMERO UN

Why Me?

You are born without your consent. As nature and nurture run their course—where will you end up in the annals of history books?

The drama that occurs inside the womb is the same drama that occurs outside of the womb: there is only room for **ONE** at the top. The good, great, and gifted will leave a legacy that is celebrated for eons into eternity.

The good, great, and gifted have to share the planet with the most vile among us: serial killers, war mongers, and jihad terrorists. Both will achieve lengthy citations in our history books.

The most cruel and cold-blooded among us also rise to the top spot: Hitler, Osama bin Laden, and Sinwar. The evil doers rise to positions of power and take advantage of the unconscious, ignorant, and irrational to lead them into the abyss of war, genocide, and death.

With no claim to fame or infamy, the ordinary majority are forgotten as if they had never lived a life.

Prodigy Sharon Esther Lampert
Genius: The Gift of Divine Revelation
EXODUS 31: 1-3

NEW SCIENTIFIC THEORY

SPERM MANIFESTO

10 RULES FOR THE ROAD

There Is Only Room for **ONE** at The Top

NUMERO UNO

NO **F**AKES!
NO **F**AT!
NO **F**LUFF!
NO **F**ILLER!
NO **F**LOPS!
NO **F**UDGE!
NO **F**BOMB!

Awesome **A**rt of **A**lliteration
Using One Letter of the **A**lphabet

What Do Books Do?
—Written in Letter **E**

8 Goalposts of Education
—Written in Letter **E**

Women Empowerment
—Written in Letter **E**

Sharon's Biography — Written in Letters **F, B**, and **P**

CUPID
Language of Love
—Written in Letter **C**

DESTINY
Are You Living Life By Default or By Design?
—Written in Letter **D**

TEMPORARY INSANITY
We Are Building Our Lives on a **S**and Trap
—Written in Letter **S**

PUBLISH: THE SECRET SAUCE OF BOOK SALES
How to Make Money Selling Books
—Written in Letter **P**

THERAPY
What **T**est Did You **T**ake **T**oday?
—Written in Letter **T**

POWER
Purgatory or **P**aradise?
—Written in Letter **P**

10 Rules for The Road

NUMERO UNO

NEW SCIENTIFIC THEORY

SPERM MANIFESTO

THERE IS ONLY ROOM FOR **ONE** AT THE TOP

By Sharon Esther Lampert

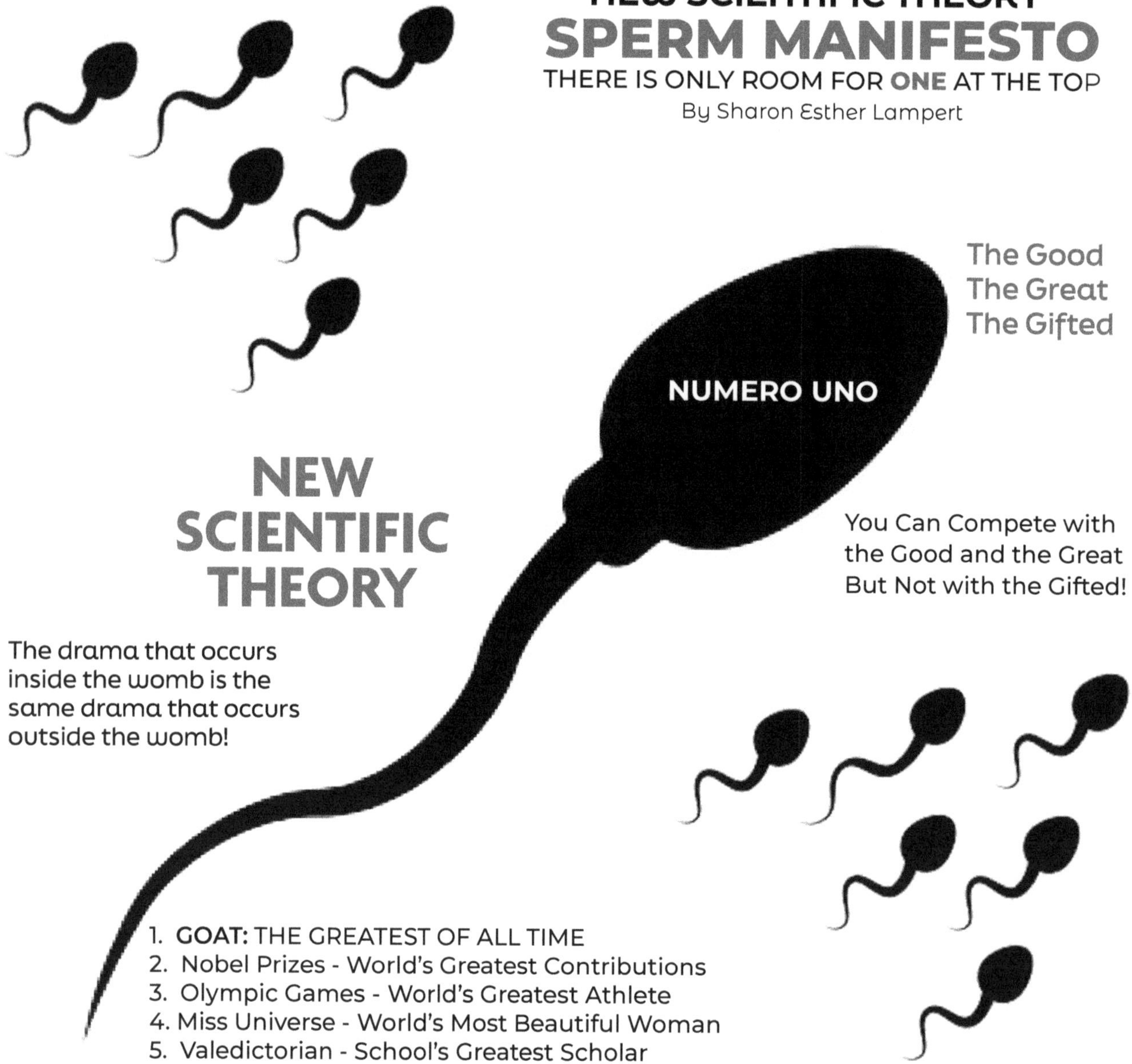

The Good
The Great
The Gifted

NEW SCIENTIFIC THEORY

You Can Compete with the Good and the Great But Not with the Gifted!

The drama that occurs inside the womb is the same drama that occurs outside the womb!

1. **GOAT:** THE GREATEST OF ALL TIME
2. Nobel Prizes - World's Greatest Contributions
3. Olympic Games - World's Greatest Athlete
4. Miss Universe - World's Most Beautiful Woman
5. Valedictorian - School's Greatest Scholar
6. Dog Show - World's Greatest Dog
7. JEWS: 1% Earn 25% Nobel Prizes and 40% of Every Prize in Every Field
8. **GOAT:** SHARON ESTHER LAMPERT: SEXIEST CREATIVE GENIUS IN HUMAN HISTORY
9. **GOAT: SCHMALTZY** World Famous Piano-Playing Cat

Introduction

SPERM MANIFESTO
THERE IS ONLY ROOM FOR ONE AT THE TOP

Part 1. NEW SCIENTIFIC THEORY!
This book is the first of its kind to elucidate a new scientific theory: **The drama that occurs inside the womb is the same drama that occurs outside the womb.**

NUMERO UNO
Similar to the sperm that competes with billions of other sperm to fertilize an egg —only one sperm will survive while the rest perish. At every stage of life, there will be only one who is chosen who will be able to climb to the apex of the mountaintop. There are many mountaintops to climb, and there is only room for **ONE** at the top:

1. **GOAT: THE GREATEST OF ALL TIME**
2. Nobel Prizes: World's Greatest Contributions
3. Olympic Games: World's Greatest Athlete
4. Miss Universe: World's Most Beautiful Woman
5. Valedictorian: School's Greatest Scholar
6. Dog Show: World's Greatest Dog
7. GOAT! Sharon Esther Lampert: Sexiest Creative Genius in Human History

GOOD, GREAT, AND GIFTED
The select minority among us will leave a legacy for future generations to inspire, emulate, and celebrate. They will leave indelible footprints in the sand that last for eons into eternity. The gifted shift paradigms and alter the course of human history.

Part 2. TEN RULES FOR THE ROAD
This book is a self-help book of insights to help you navigate life's treacherous terrain on your way to **NUMERO UNO!**

Part 3. LESSONS LEARNED: IF ONLY YOU KNEW THEN WHAT YOU KNOW NOW!

Part 4. WORKBOOK AND JOURNAL
Many people wish they could live their lives all over again. The workbook & journal will help you understand the world around you and the world inside of you.

TEN RULES FOR THE ROAD

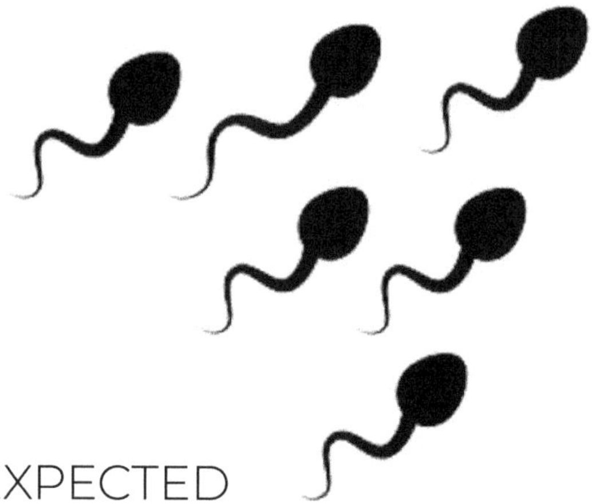

Rule 1. METAMORPHOSIS

Rule 2. THE DARK TRIAD

Rule 3. CURVEBALLS: EXPECT THE UNEXPECTED

Rule 4. LIVE YOUR TRUTH: BREAK OUT OF BIRTH BUBBLE

Rule 5. SO YESTERDAY: CHOOSE A FRESH START

Rule 6. GOD IS GO! DO! – THE 22 COMMANDMENTS (p. 59)

Rule 7. FIND THE HELPERS

Rule 8. ME WE: SELF-LOVE & BONUS LOVE

Rule 9. HAPPINESS IS AN ACT OF DEFIANCE

Rule 10. GOOD, GREAT, AND GIFTED

The two most important days in your life are the day you were born and the day you find out **WHY**?
— Mark Twain

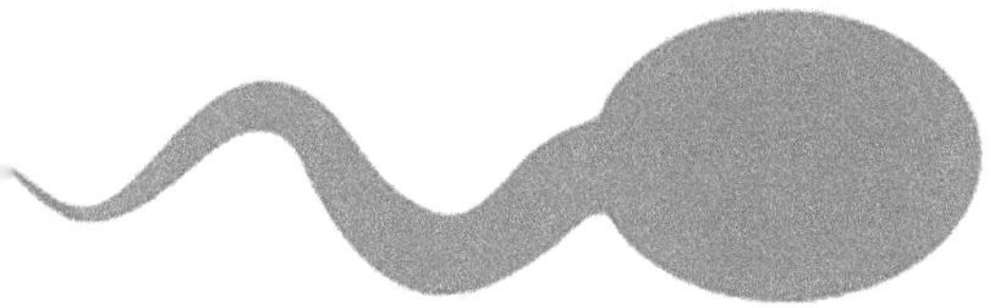

Chapter 1

METAMORPHOSIS

EVERY DAY OF YOUR LIFE YOU WILL UNDERGO METAMORPHOSIS

USE IT OR LOSE IT!

Q: Do you **remember** the 9 months inside of the womb?

1. Conception
2. Cleavage
3. Morula
4. Blastocyst
5. Implantation
6. Gastrulation
7. Organogenesis
8. Foetal

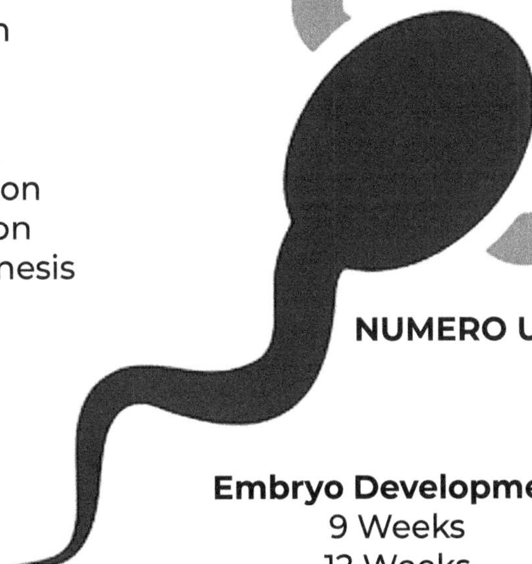

NUMERO UNO

Embryo Development
9 Weeks
12 Weeks
16 Weeks
20 Weeks
24 Weeks
28 Weeks
32 Weeks
36 Weeks
40 Weeks

Develop Your Mind, Body, and Spirit

When you die,
you will not be
able to **remember**
having lived a life!

Sharon Esther Lampert
PHILOSOPHER QUEEN

WISDOM

Transform Information into Knowledge and Knowledge into **Wisdor**

DEVELOP
MIND, BODY, AND SPIRIT

USE IT OR LOSE IT!

EVERY DAY YOU ARE MOVING FORWARD OR BACKWARD

WISDOM

FIGHT TO LIVE
LIVE TO FIGHT
BORN TO DIE

Philosopher Queen Sharon Esther Lampert

1. METAMORPHOSIS

Every Day of Your Life You Are Undergoing a Metamorphosis

Why Me? You are born without your consent. Upon conception, your nine months in the womb were a turbulent time of cell division. After birth, you will continue to undergo daily physical, mental, and spiritual changes. **MOTHER NATURE** engineered these changes—but now it is up to you to participate in the process.

As you undergo the daily process of **METAMORPHOSIS**, you are tasked with the arduous development of your mind, body, and spirit. Our minds require knowledge: **Fact Over Fiction** and **Truth Over Lie.** Our bodies require **Whole Foods, Exercise, and Sleep**—a daily delicate balance of exertion and recovery. Our spirit requires daily affirmations of **Wisdom: Light Over Darkness.**

Each and every day you are either moving forward or moving backward: For example, without daily repetition, your mind will forget what it once knew. Without daily exercise, our body will lose its muscle mass. Without daily positive spiritual affirmations, our souls wither into fear, anxiety, stress, despair, depression and the dark night of the soul. **USE IT OR LOSE IT!**

MIND: Fact Over Fiction and Truth Over Lie

BODY: Eat Whole Foods (no ingredient labels)
Avoid Processed Food
Chemicals + Inflammation = Disease

SPIRIT: Wisdom Is Light Over Darkness

THE 22 COMMANDMENTS p. 59

TEN RULES FOR THE ROAD

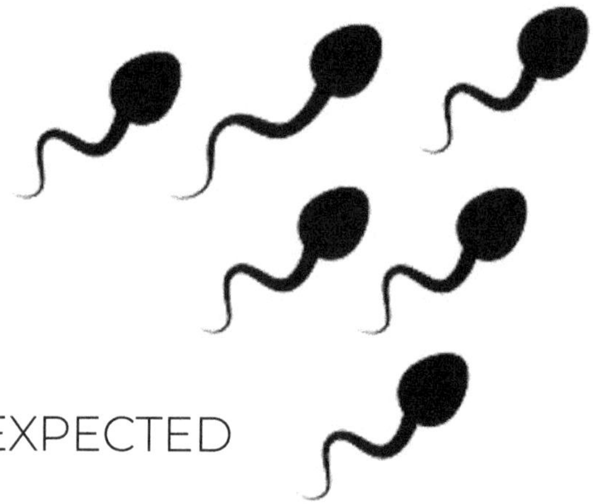

Rule 1. METAMORPHOSIS

Rule 2. THE DARK TRIAD

Rule 3. CURVEBALLS: EXPECT THE UNEXPECTED

Rule 4. LIVE YOUR TRUTH: BREAK OUT OF BIRTH BUBBLE

Rule 5. SO YESTERDAY: CHOOSE A FRESH START

Rule 6. GOD IS GO! DO! – THE 22 COMMANDMENTS (p. 59)

Rule 7. FIND THE HELPERS

Rule 8. ME WE: SELF-LOVE & BONUS LOVE

Rule 9. HAPPINESS IS AN ACT OF DEFIANCE

Rule 10. GOOD, GREAT, AND GIFTED

ONE GLOBAL ENEMY
IGNORANCE

Sharon Esther Lampert
PHILOSOPHER QUEEN

Chapter 2
THE DARK TRIAD

1. UNCONSCIOUS
2. IGNORANT
3. IRRATIONAL

BLIND LEAD THE BLIND

THE DARK TRIAD

I Am Unconscious!
I Am Ignorant!
I Am Irrational!
I Am Born This Way!
I Will Live This Way!
I Will Die This Way!

Sharon Esther Lampert
PHILOSOPHER QUEEN

NUMERO UNO

Like **MOSES,**
You Will Walk
40 YEARS
In The
DESERT
Until You Reach
ENLIGHTENMENT
10 COMMANDMENTS
And You Will
Never Enter The
PROMISED LAND

TRIAL'N'ERROR:
EVERYTHING
GOES
WRONG
BEFORE
ANYTHING
GOES
RIGHT!

IN EVERY GENERATION
THE BLIND LEAD THE BLIND

Sharon Esther Lampert
PHILOSOPHER QUEEN

WISDOM

Transform Information into Knowledge and Knowledge into **Wisdo**

DAILY PRACTICE:

1. Q: Fact or Fiction?
2. Q: Truth or Lie?
3. Q: Light or Darkness?

SEEK WISDOM

LIGHT OVER DARKNESS

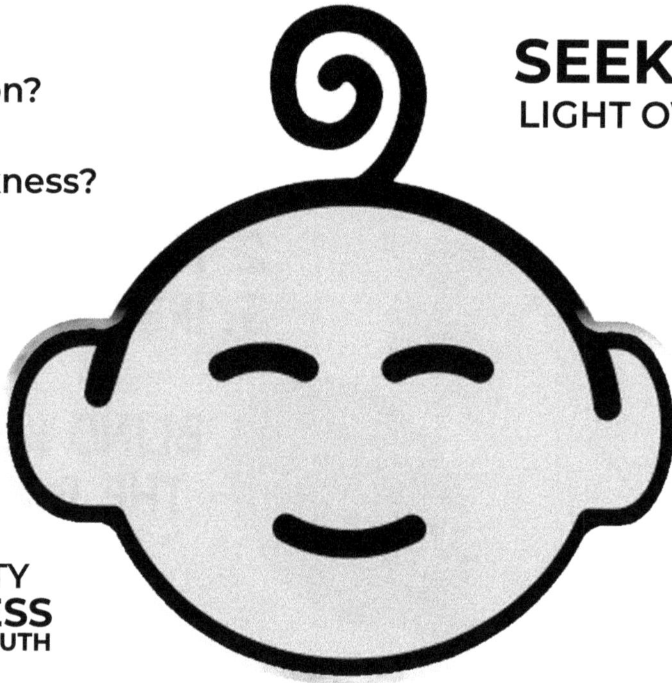

THERE IS ONLY ONE

TRUTH

NO ONE HAS THE TRUTH

Sharon Esther Lampert
PHILOSOPHER QUEEN

LAW OF
INEXTRICABILITY
TRUTHLESSNESS
TRUTH + LIE = HALF-TRUTH

WISDOM

Q: HOW SCARY IS THAT?

8 BILLION PEOPLE

BORN INTO THE DARK TRIAD

We Are Born **Unconscious, Ignorant,** and **Irrational**

We Will Live This Way! We Will Die This Way!

IN EVERY GENERATION THE BLIND LEAD THE BLIND

2. THE DARK TRIAD

Unconscious, Ignorant, Irrational

1. Billions of people have lived and died on this planet. Even if both your parents earned NOBEL PRIZES, you will be born into **THE DARK TRIAD: UNCONSCIOUS, IGNORANT, and IRRATIONAL.**

Like Moses's 40-year walk in the desert, you will walk for 40 years until you reach enlightenment: **THE 10 COMMANDMENTS.** Like Moses, you will never enter the **PROMISED LAND.**

THE DARK TRIAD
I Am Unconscious!
I Am Ignorant!
I Am Irrational!
I Am Born This Way!
I Will Live This Way!
I Will Die This Way!

2. **IN EVERY GENERATION, THE BLIND LEAD THE BLIND!**

3. **TRIAL'N'ERROR**
EVERYTHING GOES WRONG, BEFORE ANYTHING GOES RIGHT!

4. **SEEK WISDOM: LIGHT OVER DARKNESS**
Transform information into knowledge and transform knowledge into **WISDOM.** Schools do not offer a **Department of Wisdom.** You will learn the hard way through life experiences. Every day is a test!

5. **MY DAILY PRACTICE:**
Your mind must be able to discriminate and discern three criteria:
    1. Q: Fact or Fiction?
    2. Q: Truth or Lie?
    3. Q: Light or Darkness?

TEN RULES FOR THE ROAD

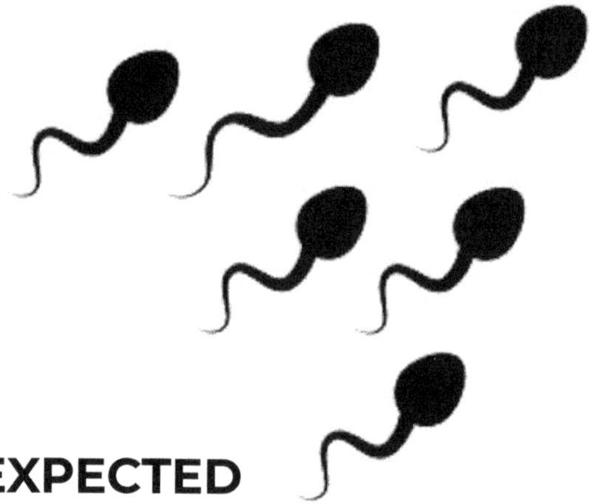

Rule 1. METAMORPHOSIS

Rule 2. THE DARK TRIAD

Rule 3. CURVEBALLS: EXPECT THE UNEXPECTED

Rule 4. LIVE YOUR TRUTH: BREAK OUT OF BIRTH BUBBLE

Rule 5. SO YESTERDAY: CHOOSE A FRESH START

Rule 6. GOD IS GO! DO! – THE 22 COMMANDMENTS (p. 59)

Rule 7. FIND THE HELPERS

Rule 8. ME WE: SELF-LOVE & BONUS LOVE

Rule 9. HAPPINESS IS AN ACT OF DEFIANCE

Rule 10. GOOD, GREAT, AND GIFTED

EVERY DAY IS A TEST!

Q: What Test Did You Take Today?

— Sharon Esther Lampert

Chapter 3

CURVEBALLS EXPECT THE UNEXPECTED

THE PHYSICAL WORLD IS A TORTURE CHAMBER OF UNSPEAKABLE HORRORS

NUMERO UNO I AM LUCKY!

or Centuries:

Domestic Dysfunctional Drama
. Natural Disasters
. Diseases
. Man's Inhumanity to Man:
a. Discrimination
b. Racism
c. Sexism
d. Religious Strife: Convert or Die!

4 Global Catastrophes of
iolence Against Women!

AFGHANISTAN
Millions of Girls
Can't Go to School

Read My Book:
SILLY LITTLE BOYS
0 RULES OF MANHOOD
How Do Silly Little Boys
row into Big Sane Men?

PTSD

POST
TRAMATIC
STRESS
SYNDROME

24 HOURS

WORLD PEACE
8 HOURS A DAY
WHEN
EVERY
LIVING
BEING IS
SLEEPING
16 HOURS OF CHAOS
DRAMA & TRAUMA

— Sharon Esther Lampert

Centuries of Islamic Jihad Terrorism
Radical Islamic Jihadists Hijacked Countries; 2024 Protestors Chant:
"Death to USA!"
"Death to Canada!"
"Death to Australia!"
"Death to Europe!"

(Check out Yemen's flag!)

"Allah is our objective. The Prophet is our leader. Qur'an is our law. Jihad is our way. Dying in the way of Allah is our highest hope."

— Muslim Brotherhood

WISDOM

Transform Information into Knowledge and Knowledge into **Wisdom**

Nothing Is Going to Hit
As Hard As Life
But It Ain't How Hard You Hit
It's About How Hard You Get Hit
and Keep Moving Forward

ROCKY BALBOA, 2006

SOME PEOPLES **DAYDREAMS** ARE **SCARIER** THAN THEIR **NIGHTMARES**

Sharon Esther Lampert
PHILOSOPHER QUEEN

MY DAILY PRACTICE:

1. TODAY I WAS LUCKY
2. TODAY I WAS UNLUCKY
3. TODAY I WAS LUCKY & UNLUCKY

WISDOM

OUR PHYSICAL WORLD IS ORGANIZED BY

15 LAWS OF INEXTRICABILITY

Physicist Sharon Esther Lampert

Read My Book: **NOW YOU KNOW EVERYTHING ABOUT EVERYTHING**

3. CURVEBALLS

We Are Unprepared for Life's Misfortunes, Mother Nature, or Man's Inhumanity

GENERATIONS OF TRAUMA: Every person you know has suffered a trauma and has a degree of **PTSD**: It is a broken world or broken people. The physical world is a torture chamber of unspeakable horrors.

CURVEBALL: GLOBAL GENOCIDE COVID19

In 2020, the entire world suffered the same **CURVEBALL: COVID19**. Millions of people of every nationality perished from the vicious invisible and intangible mutating-murdering-monster microbes. The virus turned all of us into murderers of our loved ones—not our enemies! Every interaction was infectious and calamitous. Even our houses of worship were closed to the public: **GOD WAS NOT AN ESSENTIAL SERVICE!** We barricaded ourselves inside as wild animals roamed our streets. A virus targeting every living being with the intention to exterminate all living beings is a **GENOCIDE! OY VEY!** Hugs & kisses were replaced with waves of a hand, **"Stay Away!"**

CURVEBALL: NATURAL DISASTERS

Each and every day, we bear witness to natural disasters as hurricanes, tornadoes, and earthquakes devastate our communities causing billions of dollars of damage in a couple of hours. The day after, our homes are swept away, our pets are found floating on wooded planks, and our neighbors are sitting on their rooftops waiting for emergency rescue helicopters.

CURVEBALL: MAN'S INHUMANITY TO MAN (LOSE! LOSE! Gameplan)

We are born unconscious, ignorant, and irrational. In every generation, the blind lead the blind. We are all going to die—but we engage in war to destroy each other—unable to wait for nature to take its course.

CURVEBALL: MOTHER NATURE and GLOBAL FEEDING FRENZY

Every sentient being is afraid of every sentient being: Who is lunch today?

TEN RULES FOR THE ROAD

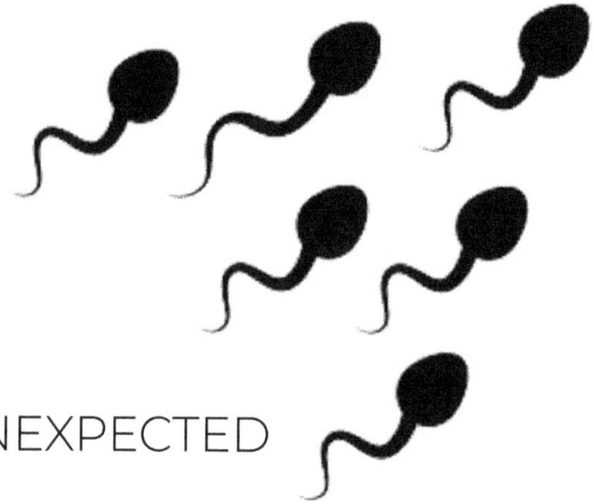

Rule 1. METAMORPHOSIS

Rule 2. THE DARK TRIAD

Rule 3. CURVEBALLS: EXPECT THE UNEXPECTED

Rule 4. LIVE YOUR TRUTH: BREAK OUT OF BIRTH BUBBLE

Rule 5. SO YESTERDAY: CHOOSE A FRESH START

Rule 6. GOD IS GO! DO! – THE 22 COMMANDMENTS (p. 59)

Rule 7. FIND THE HELPERS

Rule 8. ME WE: SELF-LOVE & BONUS LOVE

Rule 9. HAPPINESS IS AN ACT OF DEFIANCE

NUMERO UNO

Rule 10. GOOD, GREAT, AND GIFTED

Q1: What MOVIE Are You In?

EVERYONE IS IN HIS/HER OWN MOVIE
EVERY MOVIE HAS BEEN DONE!

Chapter 4
LIVE YOUR TRUTH

99% of People Live Their Lives on a DEFAULT Setting

BREAK OUT OF YOUR BIRTH BUBBLE!

NUMERO UNO

Q2: What TEST Did You Take Today?

Q3: What GAME Did You Play Today?

3 GAMES IN LIFE:

1. WIN! WIN! (cooperation)
2. WIN! LOSE! (competition)
3. LOSE! LOSE! (mutual destruction)

FIND YOUR **PASSION**
FULFILL YOUR **POTENTIAL**
AND FIND YOUR **PLACE**
IN THE WORLD

PHOTON SUPERHERO EDUCATION

SMARTGRADES
BRAIN POWER REVOLUTION

Q4: WHAT POSITION?

1. **In the Game:** (BlacknBlue)
2. **On the Bench:** (Waiting to Play)
3. **On the Sidelines:** (Rah! Rah! or Boo! Boo!)

WISDOM

Transform Information into Knowledge and Knowledge into **Wisdom**

BE THE CENTER OF YOUR OWN UNIVERSE

99% of People Wish They Could Live Their Lives All Over Again to Get It Right!

IF ONLY I KNEW THEN WHAT I KNOW NOW!

Start Life on Plan A
End Life on Plan W

STAY IN YOUR OWN LANE AND RUN YOUR OWN RACE

WISDOM

Listen to your own inner voice!

Everyone gets a different set of cards. What you do with the cards you are dealt determines your destiny!

Philosopher Queen Sharon Esther Lampert

Read My Book: DESTINY: DARE TO DREAM

4. LIVE YOUR TRUTH

Q: Are You Living Your Life By **Default** or By **Design**?

Step 1. Listen to Your Own Inner Voice!
Step 2. Break Out of Your Birth Bubble!
Step 3. Stay in Your Own Lane and Run Your Own Race!
Step 4. Be the Center of Your Own Universe!

100% OF PEOPLE LIVE THEIR LIVES ON A DEFAULT SETTING

We are born into **BIRTH BUBBLES.** Like a Floridian sea turtle, we have to break out of our birth bubble to achieve our own unique destiny. 300 baby sea turtles enter the vast ocean to swim the trecherous waves—and each turtle sets sail on his own unique adventure (YouTube video).

Everyone Gets a Different Set of Cards. What You Do with the Cards You Are Dealt in Life Determines Your Destiny!

Similarly, in school, each one of our classmates chooses a different profession that suits a unique skill set of personality, talents, and gifts. In this day and age, it is not uncommon to see friends change almost everything from their hair color, to their religion, to their partners, to their pets, to their houseplants, to their professions, to living abroad millions of miles from where they were born.

Whatever You Desire—**LIVE YOUR TRUTH**—Manifest Your Destiny

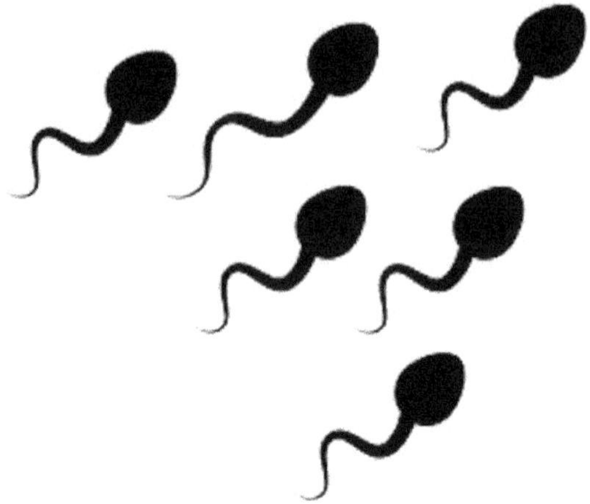

TEN RULES FOR THE ROAD

Rule 1. METAMORPHOSIS

Rule 2. THE DARK TRIAD

Rule 3. CURVEBALLS: EXPECT THE UNEXPECTED

Rule 4. LIVE YOUR TRUTH:BREAK OUT OF BIRTH BUBBLE

Rule 5. SO YESTERDAY: CHOOSE A FRESH START

Rule 6. GOD IS GO! DO! – THE 22 COMMANDMENTS (p. 59)

Rule 7. FIND THE HELPERS

Rule 8. ME WE: SELF-LOVE & BONUS LOVE

Rule 9. HAPPINESS IS AN ACT OF DEFIANCE

Rule 10. GOOD, GREAT, AND GIFTED

One Third of Your Life
SLEEPING

The richest man and the poorest man wake up every day with the same problems:

I'm hungry!
I'm thirsty!
I'm naked!

24/7 REPEAT!

Chapter 5
SO YESTERDAY!

Choose a
FRESH START
Over Righting
the Wrongs
of the Past

NUMERO UNO

CENTURIES of REPEAT!

24 HOUR CYCLE
1. Eat
2. Pee
3. Poop
4. Clean
5. Work
6. Sleep

24/7 REPEAT!

WISDOM

Transform Information into Knowledge and Knowledge into **Wisdom**

99% OF LIFE IS ON REPEAT

DAILY PRACTICE
Mindfulness, Meditation, and Inspirational Music for
INNER PEACE

Start Each Day with a
FRESH START
"SO YESTERDAY!"

Sharon Esther Lampert
PHILOSOPHER QUEEN

WISDOM

Start Each Day with the
THE SERENITY PRAYER

I didn't cause it.
I can't control it.
I can't change it.

5. SO YESTERDAY!

Every Day Choose a Fresh Start Over Righting the Wrongs of the Past

FRESH START! Approach each day anew!

Life can only be lived one day at a time and one foot in front of the other. And most of the day is on **REPEAT!**

REPEAT! 24/7 DAYS A WEEK
EAT, PEE & POOP, CLEAN, WORK, SLEEP

Most of life will be spent eating, peeing and pooping, cleaning, working, and sleeping. Redundancy is the norm!

On any given day, there is only a small window of opportunity to take a step forward to manifest a goal.

You Are Always Standing on the Shoulders of Previous Generations that Made Your Contribution Possible!

When we read the daily newspaper, it appears as if the world around us is in a state of perpetual chaos: stress and anxiety. It is!

Try not to get sucked into all of the **DRAMA** and **TRAUMA!** Most of the catastrophe is also on **REPEAT!** for **CENTURIES**! Make each day count as a step in the right direction towards living your best life. Time cannot be saved, and your time is finite! Life is a short trip around the sun; death is forevermore! Stay focused on your own desires, dreams, and destiny!

Q: Whose shoulders will you stand on to achieve your destiny?

TEN RULES FOR THE ROAD

Rule 1. METAMORPHOSIS

Rule 2. THE DARK TRIAD

Rule 3. CURVEBALLS: EXPECT THE UNEXPECTED

Rule 4. LIVE YOUR TRUTH:BREAK OUT OF BIRTH BUBBLE

Rule 5. SO YESTERDAY:CHOOSE A FRESH START

Rule 6. GOD IS GO! DO! – THE 22 COMMANDMENTS (p. 59)

Rule 7. FIND THE HELPERS

Rule 8. ME WE: SELF-LOVE & BONUS LOVE

Rule 9. HAPPINESS IS AN ACT OF DEFIANCE

Rule 10. GOOD, GREAT, AND GIFTED

Does God Talk to You Too?

Chapter 6

GOD IS GO! DO!

GOD CAN ONLY DO FOR YOU WHAT GOD CAN DO THROUGH YOU!

NUMERO UNO

PRAY
As It All
Depends
On God!

WORK
As If It All
Depends
On You!

Rachel Goldberg-Polin
Mother of Israeli Hostage
Hersch Goldberg-Polin
May His Memory Be a Blessing

Dear God,
I'm hungry ...
PLANT A SEED
GOD IS GO! DO!

Dear God,
I'm naked ... **FASHION**
GOD IS GO! DO!

Dear God,
I want to fly
like a bird ...
AIRPLANES
GOD IS GO! DO!

Dear God,
I want to swim
like a fish ...
SUBMARINES
GOD IS GO! DO!

Dear God,
I want to
go to the
moon ...
SPACESHIPS
GOD IS GO! DO!

WISDOM

Transform Information into Knowledge and Knowledge into **Wisdom**

LAW OF INEXTRICABILITY
GODLESSNESS
There Are Two Worlds:
PHYSICAL WORLD &
METAPHYSCIAL WORLD

THE
22
COMMANDMENTS

A Universal Moral
Compass
For All People
For All Religions
For All Time

All You Will Ever
Need to Know
About God

p. 59

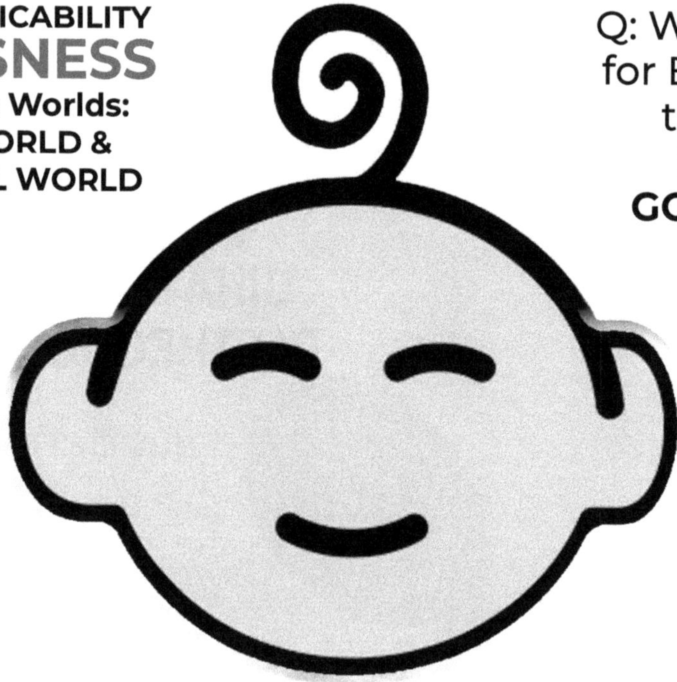

Q: Why Is It **NORMAL**
for Billions of People
to Talk to **GOD**
But **CRAZY** If
GOD Talks to Me?

Read My Book:
WHO KNEW GOD WAS SUCH A CHATTERBOX

Read My Book:
UNLEASH THE CREATOR THE GOD WITHIN
10 ESOTERIC
LAWS OF GENIUS
AND CREATIVITY

WISDOM

GOD IS GO! DO!

GOD CAN ONLY DO
FOR YOU WHAT GOD
CAN DO THROUGH YOU!
PRAY AS IF IT ALL DEPENDS ON **GOD**!
WORK AS IF IT ALL DEPENDS ON **YOU**!

Prophet Sharon Esther Lampert

6. GOD IS GO! DO!

God Can Only Do For You What God Can Do Through You

Eight billion people of diverse religious traditions pray multiple times a day to their **GOD or GODS**:

There are many religions and denominations
There are many prophets
There are many holy books
There are many holidays
There are many religious garments
There are many traditions
There are many rituals
There are many customs
But there is only one prayer:

4000 Religions: 8 Billion People of All Faiths Recite One Prayer:
"DEAR GOD: HELP ME, HEAL ME, AND PROTECT ME FROM HARM

LAW OF INEXTRICABILITY: GODLESSNESS

There are two worlds: a physical world and a metaphysical world

The **PHYSICAL WORLD** is organized by **LAWS OF INEXTRICABILITY.**

The **METAPHYSICAL WORLD:** Your mind, thoughts, and ideas are intangible and invisible entities that are beyond the scope of scientific inquiry! The **METAPHYSICAL WORLD** is governed by the **10 ESOTERIC LAWS OF GENIUS AND CREATIVITY.** This is the process of how dreamers imagine and manifest their desires and dreams into the physical world.

Pray As If It All Depends On GOD—Work As If It All Depends On **YOU!**

TEN RULES FOR THE ROAD

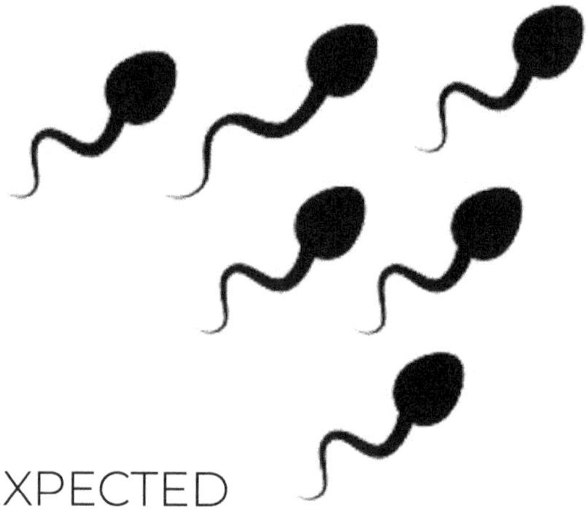

Rule 1. METAMORPHOSIS

Rule 2. THE DARK TRIAD

Rule 3. CURVEBALLS: EXPECT THE UNEXPECTED

Rule 4. LIVE YOUR TRUTH:BREAK OUT OF BIRTH BUBBLE

Rule 5. SO YESTERDAY: CHOOSE A FRESH START

Rule 6. GOD IS GO! DO! – THE 22 COMMANDMENTS (p. 59)

Rule 7. FIND THE HELPERS: A MENTOR WITH A MAP

Rule 8. ME WE: SELF-LOVE & BONUS LOVE

Rule 9. HAPPINESS IS AN ACT OF DEFIANCE

NUMERO UNO

Rule 10. GOOD, GREAT, AND GIFTED

Broken World of Broken People

5 Out of 5 People Have
Mental Health Distress

Generations of:
Drama
Trauma
Karma

YouTube is a Great Resource of Professionals Who Offer Free Advice

MOTHER NATURE

POISONOUS PLANTS
POISONOUS SNAKES
POISONOUS **PSYCHOPATHS**

Chapter 7

Find The Helpers A Mentor with a Map

Q: Do you have a support system of people who appreciate you, support you, and tell you the truth?

NUMERO UNO

5 Out of 5 People Have
Mental Health Distress

1 Out of 5 People Has an
Undiagnosed Mental Disorder

1 Out of 10 People Is a
Functional Psychotic

1 Out of 50 People
Is a **SOCIOPATH**

1 Out of 100 People
Is a **PSYCHOPATH**

EVERY STORY IS
THE SAME STORY!
CAIN & ABEL

AVOID
Naysayers!
Fearmongers!
Frenimies!
Fair-Weather Friends!

Read My Book
MONSTERS
Broken World of
Broken People

WISDOM

Transform Information into Knowledge and Knowledge into **Wisdo**

Watch Movie

A STAR IS BORN

The Climb to Stardom

(Four Movie Versions)

GOOD PEOPLE NOTHING IS A PROBLEM;
BAD PEOPLE EVERYTHING IS A PROBLEM

Sharon Esther Lampert
PHILOSOPHER QUEEN

THE DARK TRIAD
You Can't Fix Stupid!

MONSTERS
You Can't Fix Crazy!

BAD PEOPLE:
DISENGAGE
PLAY **D**UMB
PLAY **D**EAF
PLAY **D**EAD
DISAPPEAR!

Sharon Esther Lampert
PHILOSOPHER QUEEN

WISDOM

1. Make Plans: Plan A, Plan B, Plan C, Plan D ... etc.
2. Break Down Big Dreams into Achievable Steps
3. Find a Mentor Who Traveled the Same Roads
4. Get Directions: Desire, Dream, and Destiny
5. Ask for Help Along the Way

Read My Book: **DESTINY: DARE TO DREAM** —Written in Letter **D**

7. FIND THE HELPERS
Find a Mentor with a Map

Q: What Movie Are You In? Every movie has been done. Find a mentor who has traveled the same highways and get a map of directions.

SELF-HELP IS EVERYWHERE!
Our computers, cellphones, and wristwatches are computers that offer us daily guidance. Even our car's GPS computer helps us navigate the world around us. YouTube is a great resource of free help from professionals. I use it every day for just about everything:

Exercise Routine
How to Slice a Watermelon
Beauty Treatment
Medical Advice
Car Maintenance
How to Deal with a Narcissist (very fancy word for a-hole)
Taxes

SELF-HELP FOR ANNUAL COMPUTER UPGRADES
Every year, there are APPLE computer upgrades and ADOBE software upgrades. I want to know about the upgrades before I click upgrade on my APPLE computer, iphone, and wristwatch. I want to be able to use the upgrade. Watching a video takes a couple of hours each and every year, but it is worth it! Technology changes every second of every day!

SELF-HELP LIBRARY BOOKS
Before the advent of social media and the digital revolution, we were all dependent upon library books for free resources of self-help. I still use the library to take out self-help books. I have also published a few self-help books just like the one you are now reading.

TEN RULES FOR THE ROAD

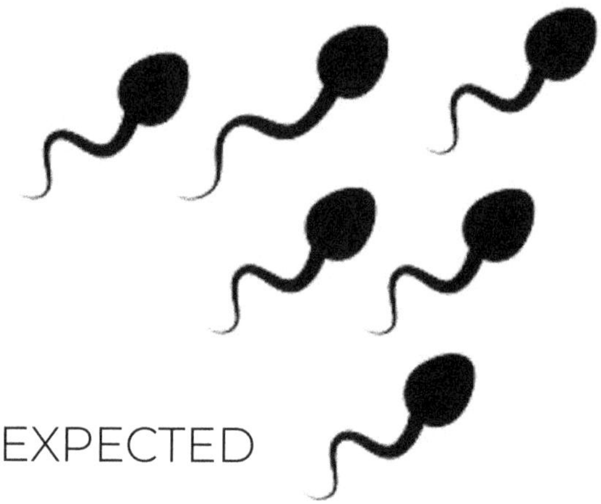

Rule 1. METAMORPHOSIS

Rule 2. THE DARK TRIAD

Rule 3. CURVEBALLS: EXPECT THE UNEXPECTED

Rule 4. LIVE YOUR TRUTH:BREAK OUT OF BIRTH BUBBLE

Rule 5. SO YESTERDAY: CHOOSE A FRESH START

Rule 6. GOD IS GO! DO! – THE 22 COMMANDMENTS (p. 59)

Rule 7. FIND THE HELPERS

Rule 8. ME WE: SELF-LOVE & BONUS LOVE

Rule 9. HAPPINESS IS AN ACT OF DEFIANCE

Rule 10. GOOD, GREAT, AND GIFTED

Chapter 8
ME WE

TRUE LOVE IS SELF-LOVE

LOVE FROM OUTSIDE IS BONUS LOVE

NUMERO UNO

Q: WHAT IS LOVE?

YOU DON'T FIND LOVE
YOU CREATE LOVE

LOVE IS A PRACTICE!
RESPECT, KINDNESS, EMPATHY

Sharon Esther Lampert
PHILOSOPHER QUEEN

If You Practice SELF-LOVE You Will Never Spend a Day in THERAPY

Sharon Esther Lampert
PHILOSOPHER QUEEN

THERE IS NO SUCH THING AS TOO MUCH LOVE

Sharon Esther Lampert
PHILOSOPHER QUEEN

BE YOUR OWN BEST FRIEND

WISDOM

Transform Information into Knowledge and Knowledge into **Wisdo**

Read My 4 Books on Relationship

1. LOVE YOU MORE THAN YESTERDA
14 Relationship Strategies
for **HAPPILY EVER AFTER**

2. CUPID — Written in Letter C

3. PHILOSOPHY OF LOVE
IF YOU PRACTICE SELF-LOVE
YOU WILL NEVER SPEND A DAY
IN THERAPY

4. SEX ON A PLATE
FOOD AS FOREPLAY

WISDOM

1. You Don't Find LOVE, You Create LOVE
2. LOVE Is a Practice: Kindness, Respect, and Empathy
3. TRUE LOVE **IS** SELF-LOVE
4. The Most Important Relationship Is the One You Have with Yourself
5. If You Practice SELF-LOVE You Will Never Spend a Day in Therapy
6. You Can Never Know Another Person: LOVE IS BLIND!
7. LOVE From Outside Yourself Is BONUS LOVE
8. LAW OF INEXTRICABILITY: LOVELESSNESS (100% of Relationships)
9. 99% of Relationships Are Conditional-Transactional Relationships
10. METAMORPHOSIS: Grow Together or Grow Apart? (chapter 1)
11. DARK TRIAD: IGNORANT, UNCONSCIOUS, IRRATIONAL (chapter 2)

8. ME WE

Q: What Is LOVE?

1. "You Don't Find LOVE, You Create LOVE" — Sharon Esther Lampert, Philosopher Queen

2. LOVE Is a Practice: Respect, Kindness, Patience, and Empathy

3. ME and WE: **TWO KINDS OF RELATIONSHIPS**
    1. **ME:** The relationship you have with yourself
    2. **WE:** The relationship you have with another person

4. ME
    1. **TRUE LOVE IS SELF-LOVE—BE YOUR OWN BEST FRIEND!**
    2. The most important relationship you have is the one you have with yourself
    3. If you practice SELF-LOVE you will never spend a day in therapy

5. WE
    1. You can never know another person: LOVE IS BLIND!
    2. Two people undergo the process of **METAMORPHOSIS:** Grow Together or Grow Apart?
    3. Two people are born into **THE DARK TRIAD:** Unconscious, Ignorant, and Irrational, Blind Lead Blind
    4. "People come into your life for a reason, a season or a lifetime" by Maya Angelou
    5. 99% of people come into your life for a reason (your parents for a season—your pets for a lifetime)
    6. **LAW OF INEXTRICABILITY: LOVELESSNESS**
        a. All people **HELP** with their strengths & **HURT** you with their weaknesses
        b. 100% of relationships are **LOVE-HATE** relationships
    7. Most people do not have enough love inside of themselves to love themselves—let alone to love you!
    8. LOVE from outside yourself is **BONUS LOVE**

6. WE: **TWO KINDS OF LOVE RELATIONSHIPS**
   Unconditional TRUE LOVE or Conditional Transactional Love

   1% Unconditional TRUE LOVE
   1. **TRUE LOVE** Is Real—But Rare—Like Winning the Lottery!
   2. Relationship Is Eternal Until Death Do You Part

   99% Conditional-Transactional Love
   1. Love Only What They Want From You
   2. Relationship Has an Expiration Date

TEN RULES FOR THE ROAD

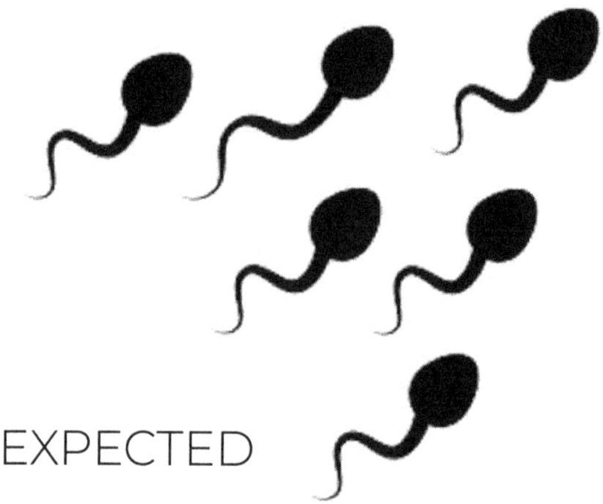

Rule 1. METAMORPHOSIS

Rule 2. THE DARK TRIAD

Rule 3. CURVEBALLS: EXPECT THE UNEXPECTED

Rule 4. LIVE YOUR TRUTH:BREAK OUT OF BIRTH BUBBLE

Rule 5. SO YESTERDAY:CHOOSE A FRESH START

Rule 6. GOD IS GO! DO! – THE 22 COMMANDMENTS (p. 59)

Rule 7. FIND THE HELPERS

Rule 8. ME WE: SELF-LOVE & BONUS LOVE

Rule 9. HAPPINESS IS AN ACT OF DEFIANCE

Rule 10. GOOD, GREAT, AND GIFTED

Most of Life Is Out of Your Control

Chapter 9

HAPPINESS IS AN ACT OF DEFIANCE

FIND THE LIGHT AND LIVE IN THE LIGHT

NUMERO UNO

BILLIONS OF PEOPLE HAVE
VED & DIED ON PLANET EARTH

Q1: Why Is Planet Earth Still a Torture Chamber of Unspeakable Horrors?

Q2: Is Life a Gift or a Punishment?

Q3: Measure the Pain?
(Life Is a Punishment)

Q4: Measure the Pleasure?
(Life Is a Gift)

A: LAWS OF INEXTRICABILITY

WISDOM

Transform Information into Knowledge and Knowledge into **Wisdo**

WISDOM

Practice Gratitude Over Grievances

Count your blessings instead of your crosses;
Count your gains instead of your losses.
Count your joys instead of your woes;
Count your friends instead of your foes.
Count your smiles instead of your tears;
Count your courage instead of your fears.
Count your full years instead of your lean;
Count your kind deeds instead of your mean.
Count your health instead of your wealth.

Unknown

9. Happiness Is an Act of Defiance

Q: Is Life a Gift or a Punishment?
Q: Measure the Pain? Q: Measure the Pleasure?

MOST OF LIFE IS OUT OF YOUR CONTROL

When you are born; Where you are born
Who are your parents; What you look like
What is your name; What is your religion
How you were educated in childhood
What language you speak
What foods you ate in childhood
How you will die; When you will die
You cannot control time (p. 50)

TRAGEDY: For many people, **DAYDREAMS** are SCARIER than **NIGHTMARES:** broken family, poverty, illiteracy, and war dominate their lives.

TIME is our most valuable asset. We have to learn how to use it and not squander it. The clock is always running ... and time cannot be saved. We cannot control time, but we can learn how to manage it to achieve our desires, dreams, and destiny. See p. 50 **99% of LIFE IS REPEAT!**

HAPPINESS: We all wish to be happy. What is happiness? Are some people born happy based on a genetic disposition? Too many of us suffer from anxiety, stress, and depression. Happiness seems elusive—5 out of 5 people suffer mental health distress. Cultivate happiness.

GRATITUDE: If you practice gratitude—you will be happier! If you hold on to the positive and let go of the negative—you will be happier!

MINDFULNESS: If you practice mindfulness, the daily practice of living in the present moment, and stop ruminating about the past and worrying about the future—you will be happier!

TEN RULES FOR THE ROAD

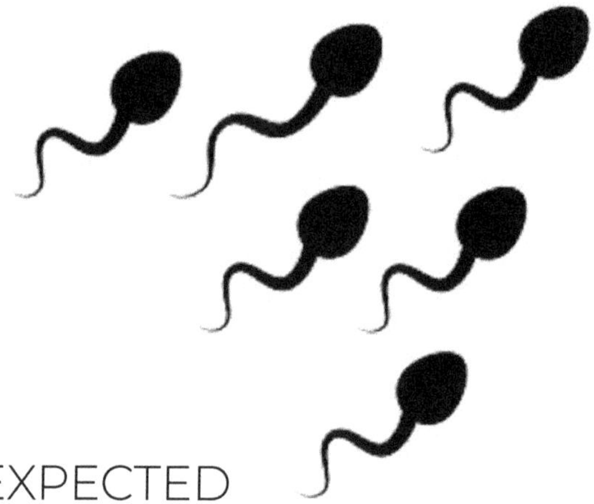

Rule 1. METAMORPHOSIS

Rule 2. THE DARK TRIAD

Rule 3. CURVEBALLS: EXPECT THE UNEXPECTED

Rule 4. LIVE YOUR TRUTH:BREAK OUT OF BIRTH BUBBLE

Rule 5. SO YESTERDAY:CHOOSE A FRESH START

Rule 6. GOD IS GO! DO! – THE 22 COMMANDMENTS (p. 59)

Rule 7. FIND THE HELPERS

Rule 8. ME WE: SELF-LOVE & BONUS LOVE

Rule 9. HAPPINESS IS AN ACT OF DEFIANCE

Rule 10. GOOD, GREAT, AND GIFTED

SPERM MANIFESTO

There Is Only Room For ONE at The Top

Chapter 10

GOOD, GREAT, AND GIFTED

You Can Compete with the GOOD and GREAT but not with the GIFTED

NUMERO UNO

1%

THE MINORITY

Leave a Legacy for Future Generations to Inspire, Emulate, and Celebrate for All Eternity

Everyone Makes Mistakes; Whoever Makes Fewer Mistakes WINS!

99%

THE MAJORITY

Will Be Forgotten As If They Were Never Born

MAY THE FORCE BE WITH YOU

YODA

Forget the MISTAKE! Remember the LESSON!

WISDOM

Transform Information into Knowledge and Knowledge into **Wisdor**

**Mistakes
May Lead to
Breakthroughs
and
Discoveries
— and —
Disruption
and Innovation
— and —
Contribution to
Civilization
— and —
Fame and Fortune
— and —
Legacy for Eons
into Eternity**

**Sharon Esther Lampert
PHILOSOPHER QUEEN**

WISDOM

**Everyone Makes Mistakes...
Forget the Mistake, Remember the Lesson
Whoever Makes Fewer Mistakes WINS!**

Philosopher Queen Sharon Esther Lampert

10. Good, Great, Gifted

Everyone Makes Mistakes
Whoever Makes Fewer Mistakes WINS!

Winners make mistakes; Losers make more mistakes. Forget the mistake —remember the lesson!

WINNERS NEVER QUIT
QUITTERS NEVER WIN
VINCE LOMBARDI

Winners had monumental setbacks—but rose from the ashes like a phoenix. Inside of our mistakes are important lessons to be learned and mastered—and can propel gigantic breakthroughs.

Successor generations stand on the shoulders of previous generations to achieve breakthroughs in science, medicine, and technology.

Artistic Gifts Are Inherited—The Gifted Have an Unfair Advantage;
You Can Compete with the Good and the Great—But Not with the Gifted!

Gifts do not come with an instructional manuel. It is up to each of us to nurture and cultivate our own talents and gifts.

You Don't Choose Art; Art Chooses You!

The **Good, Great, and Gifted** are a minority who combine raw talent & hard work. No one gets a free pass! Hard work is a vital necessity. The minority will make contributions to our civilization that will last for all eternity. We will celebrate their achievements in every generation. The majority will be forgotten as if they were never born. Fortunately, the contributions of the hardworking, talented, and gifted minority benefit the majority—KOL HAKAVOD! (all the honor)

11. WISDOM

Lessons Learned

Transform Information into Knowledge and Knowledge into WISDOM

1. Every Day of Your Life is Lived in a State of METAMORPHOSIS
2. DARK TRIAD: You Are Born **UNCONSCIOUS, IGNORANT, and IRRATIONAL**
3. **IN EVERY GENERATION THE BLIND LEAD THE BLIND**
4. Transform Information into Knowledge and Knowledge into WISDOM
5. Daily Practice: Fact or Fiction? Truth or Lie? and Light or Darkness?
6. Most of Life Is Out of Your Control—Especially the Cards You Are Dealt in Life!
7. 100% of People Live Their Lives on a **DEFAULT** Setting—Break Out of Birth Bubble!
8. Daily Practice: Q1: What **MOVIE** are you in?
Q2: What **TEST** did you take today?
Q3: What **GAME** did you play today?
Q4: What **POSITION** you you prefer?
9. 3 Games in Life: WIN! WIN! WIN! **LOSE!** and **LOSE! LOSE!**
   1. Collaboration WIN! WIN!
   2. Competition WIN! LOSE!
   3. Mutual Destruction LOSE! LOSE!
10. Listen to Your Inner Voice. Everyone Gets a Different Set of Cards.
What You Do with the Cards You Are Dealt Determines Your **DESTINY!**
11. Every Day Choose a **FRESH START** Over Righting the Wrongs of the Past
12. Our Planet is Organized by **LAWS OF INEXTRICABILITY**
13. **GOD IS GO! DO! GOD CAN ONLY DO FOR YOU WHAT GOD CAN DO THROUGH YOU**
14. **CURVEBALLS: EXPECT THE UNEXPECTED**
Planet Earth Is a Torture Chamber of Unspeakable Horrors
e.g., Natural Disasters, Diseases, and Man's Inhumanity to Man
Daily Practice: Lucky? Unlucky? Lucky and Unlucky?

15. 8 Billion People of All Faiths Recite One Prayer:
    "Dear God: Help Me, Heal Me, and Protect Me From Harm!"

16. ME and WE: **SELF-LOVE** and **BONUS LOVE**

1. Q: What Is Love? A: "You Don't Find Love, You Create Love" SEL
2. Daily Practice: Respect, Kindness, Empathy, and Patience
3. **TRUE LOVE IS SELF-LOVE**
4. Love From Outside Yourself Is **BONUS LOVE!**

ME: **SELF-LOVE**

1. The Most Important Relationship Is the One You Have with Yourself
2. If You Practice **SELF-LOVE**, You Will Never Spend a Day in Therapy!

WE: BONUS LOVE

1. You Can Never Know Another Person—**LOVE IS BLIND!**
2. Two Loves: Unconditional True Love or Conditional Transactional Love
3. 1% **TRUE LOVE** Is Real—But Rare—Like Winning a Lottery Ticket
4. 99% Conditional Transactional Love—Love Only Want They Want From You
5. **LAW OF INEXTRICABILITY: LOVELESSNESS**
   All People **HELP** You with Their Strengths and **HURT** You with Their Weaknesses
   100% of Relationships Are **LOVE-HATE** Relationships

17. **HAPPINESS IS AN ACT OF DEFIANCE**
    Daily Practice: **THE 22 COMMANDMENTS**, p. 59
    Daily Practice: **SELF-LOVE**, Gratitude Over Grievances, and FRESH START

18. **EVERYONE MAKES MISTAKES. WHOEVER MAKES FEWER MISTAKES WINS!**
    99% of People Wish They Could Live Their Lives All Over Again to Get It Right!
    IF ONLY I KNEW THEN WHAT I KNOW NOW! Mistakes May Lead to Discoveries!

19. **SPERM MANIFESTO**: There Is Only Room for **ONE** at The Top
    You Can Compete with the Good and Great—But Not with the Gifted!
    The Good, Great, and Gifted Leave a Legacy for Future Generations!

12. Workbook & Journal

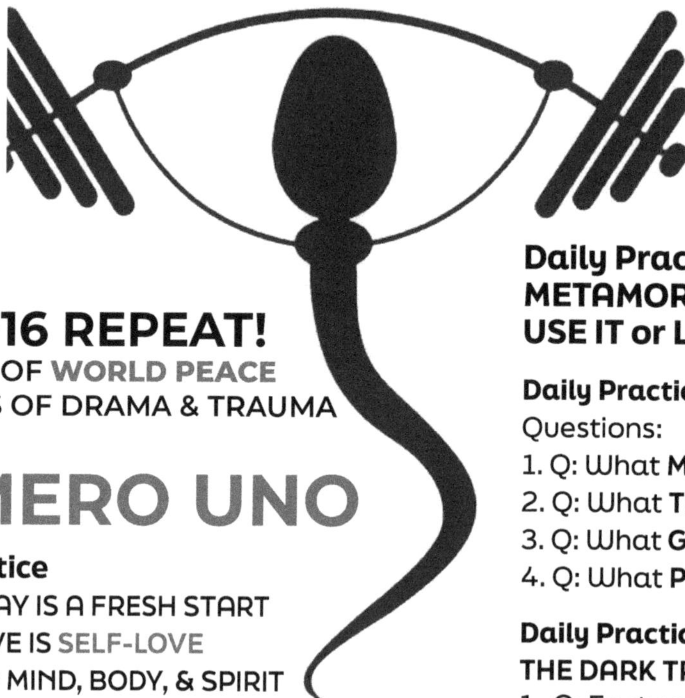

24: 8+16 REPEAT!

8 HOURS OF WORLD PEACE
16 HOURS OF DRAMA & TRAUMA

NUMERO UNO

Daily Practice

1. EVERY DAY IS A FRESH START
2. TRUE LOVE IS SELF-LOVE
3. DEVELOP MIND, BODY, & SPIRIT
4. MANAGE YOUR TIME, p. 50
5. PLAY YOUR CARDS
6. LIVE YOUR TRUTH
7. FIND THE HELPERS
8. STAND ON SHOULDERS OF PREVIOUS GENERATIONS
9. FOLLOW YOUR PASSION, FULFILL YOUR POTENTIAL, and FIND YOUR PLACE IN THE WORLD

Daily Practice
METAMORPHOSIS
USE IT or LOSE IT!

Daily Practice
Questions:

1. Q: What **MOVIE** are you in?
2. Q: What **TEST** did you take?
3. Q: What **GAME** did you play?
4. Q: What **POSITION** do you prefer?

Daily Practice
THE DARK TRIAD

1. Q: Fact or Fiction?
2. Q: Truth or Lie?
3. Q: Light or Darkness?

Daily Practice
CURVEBALLS: EXPECT THE UNEXPECTED

1. Q: Lucky?
2. Q: Unlucky?
3. Q: Lucky and Unlucky?

USE IT OR LOSE IT!

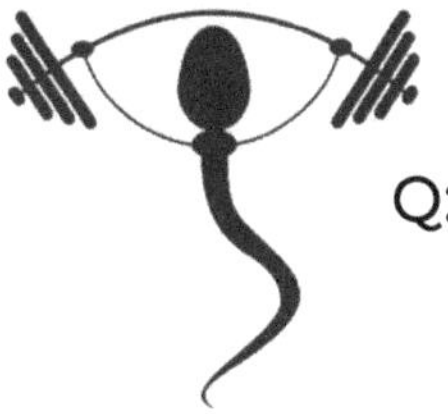

Chapter 1. Metamorphosis

Q: Are you developing your mind, body, and spirit?

Q: **USE IT OR LOSE IT?**

Cite examples:

Chapter 2. The Dark Triad

Q: Did you transform information into knowledge and knowledge into **WISDOM**?

Cite examples:

IF ONLY YOU KNEW THEN WHAT YOU KNOW NOW

Q: What plan are you on? Plan A? Plan B? ... Plan W?

Cite examples:

Chapter 3. Curveballs

Q: How many **CURVEBALLS** did you survive?

Q: Are you **Lucky**? **Unlucky**? **Lucky and Unlucky**?

Cite examples:

Read My Books:

1. THERAPY: WHAT TEST DID YOU TAKE TODAY? —Written in Letter T
2. NOW YOU KNOW EVERYTHING ABOUT EVERYTHING
   OUR WORLD IS ORGANIZED BY LAWS OF INEXTRICABILITY
3. TEMPORARY INSANITY: WE ARE BUILDING OUR LIVES ON A SAND TRAP

Chapter 4. Live Your Truth

Break Out of Your Birth Bubble

1

BIRTH BUBBLE

Q1: Do you have plans to break out of your birth bubble and chart your own course?

2

LIFE BY DEFAULT OR DESIGN

Q2: Are you living your life by **DEFAULT** or by **DESIGN**?

3

EVERYONE IS IN HIS/HER OWN MOVIE

Q3: What movie are you in?

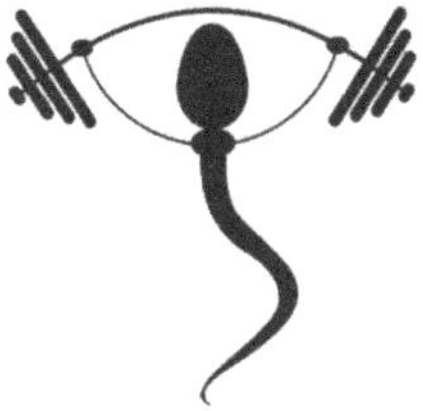

4
EVERY DAY YOU WILL TAKE A TEST
Q4: What **TEST** did you take today?

5
EVERY DAY YOU WILL PLAY A GAME
Q5: What **GAME** did you play today?

WIN:WIN WIN:LOSE LOSE:LOSE
Collaboration, Competition, or Mutual Destruction?

6
3 POSITIONS IN LIFE
Q6: What position in life do you prefer?

In the Game:	On the Bench:	In the Stands:
BlackNBlue	Waiting to Play	Rah! Rah! or Boo! Boo!

Read My Books:
The Awesome Art of Alliteration Using One Letter of the Alphabet
DESTINY: DARE TO DREAM —Written in Letter D
POWER: PARADISE OR PURGATORY? —Written in Letter P
THERAPY: EVERY DAY YOU WILL TAKE A TEST? —Written in Letter T

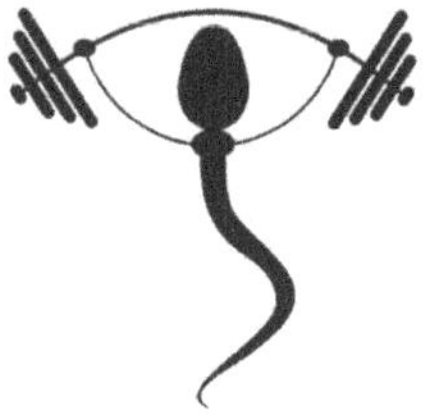

Chapter 5. So Yesterday

Q: Are you starting each day with a **FRESH START**?

1. ENERGIZE	**WIN!**
2. ORGANIZE	**WIN!**
3. PRIORITIZE	**WIN!**
4. PROJECT MANAGEMENT	**WIN!**
5. FINISH LINE	**WIN!**

7 Shifts Time Management: **WIN! W**hat's **I**mportant **N**ow!

5-9 A.M. Early Bird **WIN!**

9-12 P.M. **WIN!**

12-3 P.M. **WIN!**

3-6 P.M. **WIN!**

6-9 P.M. **WIN!**

9-12 P.M. **WIN!**

12-3 P.M. Night Owl **WIN!**

Chapter 6. GOD IS GO! DO!

PRAY As If It All Depends on **GOD**!
WORK As If It All Depends on **YOU**!

Q: Do you **WORK** as if it all depends on **YOU**?

Cite examples:

Read My Books:

1. WHO KNEW **GOD** WAS SUCH A CHATTERBOX
2. THE 22 COMMANDMENTS: ALL YOU WILL EVER NEED TO KNOW ABOUT **GOD**
3. UNLEASH THE CREATOR THE **GOD** WITHIN: 10 ESOTERIC LAWS OF GENIUS & CREATIVITY
4. **GOD** OF WHAT? LAWS OF INEXTRICABILITY

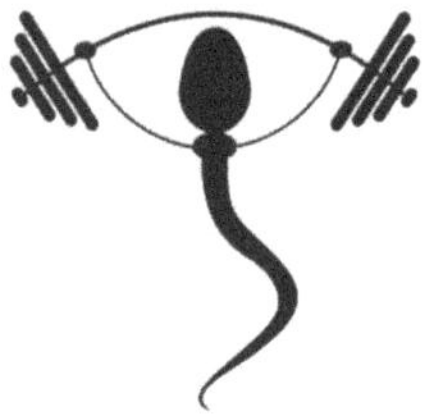

Chapter 7. Find The Helpers

Q: Do you have a support system of people who appreciate you, support you, and tell you the truth?

Cite examples:

Chapter 8. Me and We

Q: Do you practice **SELF-LOVE**? Cite examples:

Q: Did you receive any **BONUS LOVE** today? Cite examples:

Q: 99% of people are in your life for a reason? Cite examples:

Q: 99% of relationships are conditional-transactional relationships? Cite examples:

Read My Books on Love:

1. Philosophy of Love
2. CUPID: Language of Love—Written in Letter C
3. LYMTY: 14 Relationship Strategies for HAPPILY EVER AFTER
4. SEX ON A PLATE: FOOD AS FOREPLAY

Chapter 9. Happiness Is An Act of Defiance

Q: Are you practicing **SELF-LOVE?**
Q: Are you practicing **MINDFULNESS?**
Q: Are you practicing **FRESH START**?
Q: Are you practicing **Gratitude Over Grievances**?

Cite examples:

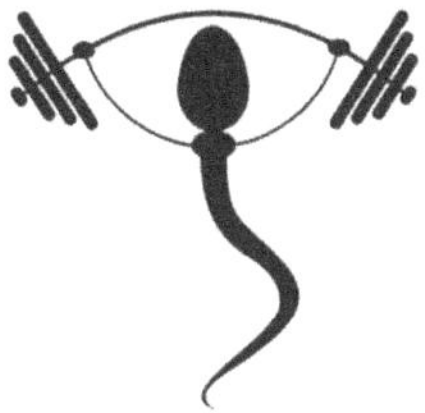

Chapter 10. Good, Great, and Gifted

Everyone Gets a Different Set of Cards

What You Do with the Cards You Are Dealt Determines Your Destiny

Q: What Cards Were You Dealt in Life?

Cite examples:

Read My Books:

1. UNLEASH THE CREATOR THE GOD WITHIN: 10 ESOTERIC LAWS OF GENIUS AND CREATIVITY
2. DESTINY: DARE TO DREAM —Written in Letter D
3. POWER: Paradise or Purgatory? —Written in Letter P

WORLD FAMOUS POEM

BE BORN

Be Born
Become Educated
Love Your Work
Make a Meaningful Contribution
to Yourself, Your Family, and Humanity
Be a True Friend to Yourself First
Have Sex with Someone You Love
Make Love with Complete Abandon
Enjoy Unconditional Love from Your Devoted Pet
Make Time to Read the Funnies and Laugh
Save Enough Money to Visit the Popular,
Pretty, and Peaceful Places of the World
Read Great Literature, Listen to Great Music
See Great Art, Watch the Great Movies
Play the Fun Sports, Dance till Dawn
Taste the Great Culinary Delights of the World
Eat Slowly, Enjoy Every Bite, and Stay in Shape
Plan One Great Adventure and Stick to the Plan
Grow Old and Wise. Leave Your Money to
Someone You Love—Who Loves You Back
Die in Your Sleep

Sharon Esther Lampert
SEE THE WORLD THROUGH THE EYES OF A CREATIVE GENIUS

What Happens When You Dress Up Albert Einstein as Marilyn Monroe?

SHARON ESTHER LAMPERT

PHENOMENON

SCIENTIST, ARTIST, EDUCATOR, THEOLOGIAN

Published: 80 BOOKS - Answered: 50 Questions

Filmmaker: 50+ "One Nominated Film"

PRODIGY
UNLEASH THE CREATOR
THE GOD WITHIN
10 ESOTERIC LAWS OF
GENIUS AND CREATIVITY

PROPHET
WHO KNEW GOD WAS
SUCH A CHATTERBOX
GOD IS GO! DO!
THE 22 COMMANDMENTS
A UNIVERSAL MORAL COMPASS

PHYSICIST
NOW YOU KNOW
EVERYTHING ABOUT
EVERYTHING: LAWS
OF INEXTRICABILITY

PSYCHOBIOLOGIST
SPERM MANIFESTO
THERE IS ONLY ROOM
FOR ONE AT THE TOP

PHILOSOPHER QUEEN

WOMEN HAVE
ALL THE POWER

THE PHILOSOPHY
OF LOVE

THE PHILOSOPHY
OF EVIL: THE
DOUBLE WHAMMY

PRODUCER
80+ Short Films

PSYCHIATRIST
14 Relationship Strategies

POET (25+ Books)
WORLD POETRY RECORD

THE GREATEST POEMS
EVER WRITTEN ON
EXTRAORDINARY
WORLD EVENTS

THE AWESOME ART OF
ALLITERATION USING
ONE LETTER OF THE
ALPHABET (6 Books)

FIRST WOMAN TO WRITE
BOOK ON JEWISH HISTORY
IN 5 MINUTES—4 REFRAINS

#1 POETRY WEBSITE FOR
STUDENT PROJECTS

PALADIN OF EDUCATION (25+ Books)

SMARTGRADES
BRAIN POWER
REVOLUTION

40 UNIVERSAL
GOLD STANDARDS
OF EDUCATION

HOW DOES
LEARNING
TAKE PLACE?

8 GOALPOSTS OF
EDUCATION

PINUP
SEXIEST CREATIVE
GENIUS IN
HUMAN HISTORY

GENIUS: EXODUS 31: 1-3
THE GIFT OF DIVINE REVELATION

www.SharonEstherLampert.com
FANS@SharonEstherLampert.com

What Happens When You Dress Up Albert Einstein As Marilyn Monroe

SHARON ESTHER LAMPERT

Prodigy: **10 LAWS OF GENIUS AND CREATIVIT**
Prophet: GOD IS GO! DO! 22 COMMANDMENTS
Physicist: **LAWS OF INEXTRICABILITY**
Psychobiologist: **SPERM MANIFESTO**
Philosopher: WOMEN HAVE ALL THE POWER
Peacemaker: WORLD PEACE EQUATION
Poet: **POETRY WORLD RECORD**
Paladin of Education: SMARTGRADES
PHOTON SUPERHERO OF EDUCATION
Princess Kadimah: 8TH PROPHETESS OF ISRA
Producer: Filmmaker
President
Publisher
Piano-Playing Cat (YouTube video)
Player: NYU Varsity B-Ball Team, Center
Performer: Vocalist (YouTube video)
Playdate
Painter
Photograper
Princess & Pea
Phoenix
PINUP

NYU
Honored Sharon Lampert with an Award for **"Multi-Interdisciplinary Studies"**
(YouTube video)

WEBSITES
SharonEstherLampert.com
WorldFamousPoems.com
Schmaltzy.com
TrueLoveBurnsEternal.com
WritersRunTheWorld.com
PalmBeachBookPublisher.com
WomenHaveAllThePower.com

EDUCATION
Smartgrades.com
BooksNotBombs.com

YOU HAD TO OUTDO MOSES

THE 22 COMMANDMENTS

All You Will Ever Need to Know About God
A Universal Moral Compass
For All People, For All Religions, For All Time

1. **LIFE** Over Death
2. **STRENGTH** Over Weakness
3. **DEED** Over Sin
4. **LOVE** Over Hatred
5. **TRUTH** Over Lie
6. **COURAGE** Over Fear
7. **OPTIMISM** Over Pessimism
8. **SHARING** Over Selfishness
9. **PRAISE** Over Criticism
10. **LOYALTY** Over Abandonment
11. **RESPONSIBILITY** Over Blame
12. **GRATITUDE** Over Envy
13. **REWARD** Over Punishment
14. **ALLIES** Over Enemies
15. **CREATION** Over Destruction
16. **EDUCATION** Over Ignorance
17. **COOPERATION** Over Competition
18. **FREEDOM** Over Oppression
19. **COMPASSION** Over Indifference
20. **FORGIVENESS** Over Revenge
21. **PEACE** Over War
22. **JOY** Over Suffering

By Sharon Esther Lampert
Princess Kadimah, 8TH Prophetess of Israel
Read My Books: **THE 22 COMMANDMENTS and**
WHO KNEW GOD WAS SUCH A CHATTERBOX

GENIUS: THE GIFT OF DIVINE REVELATION

Published 80+ Books

Prodigy—Princess & Pea!

UNLEASH THE CREATOR THE GOD WITHIN: 10 ESOTERIC LAWS OF GENIUS AND CREATIVITY
AWESOME ART OF ALLITERATION USING ONE LETTER OF THE ALPHABET (6 Books)

Prophet

WHO KNEW GOD WAS SUCH A CHATTERBOX: A WORKING DEFINITION OF GOD—GOD IS GO! DO!
22 COMMANDMENTS: ALL YOU WILL EVER NEED TO KNOW ABOUT GOD

Physicist

NOW YOU KNOW EVERYTHING ABOUT EVERYTHING: LAWS OF INEXTRICABILITY

Psychobiologist

SPERM MANIFESTO: THERE IS ONLY ROOM FOR ONE AT THE TOP - 10 RULES FOR THE ROAD

Philosopher Queen

WOMEN HAVE ALL THE POWER BUT HAVE NEVER LEARNED HOW TO USE IT
THE PHILOSOPHY OF LOVE
THE PHILOSOPHY OF EVIL

Paladin of Education (+25 Books)

PHOTON SUPERHERO OF EDUCATION

SMARTGRADES BRAIN POWER REVOLUTION

- THE SILENT CRISIS DESTROYING AMERICA'S BRIGHTEST MINDS—BOOK OF THE MONTH
- EVERY DAY AN EASY A!
- 40 UNIVERSAL GOLD STANDARDS OF EDUCATION
- HOW DOES LEARNING TAKE PLACE

Poet (+25 Books)

ONE OF THE WORLD'S GREATEST POETS
POETRY WORLD RECORD: 120 WORDS OF RHYME
THE GREATEST POEMS EVER WRITTEN ON EXTRAORDINARY WORLD EVENTS
HTTP://FAMOUSPOETSANDPOEMS.COM/POETS.HTML

Princess Kadimah

8TH PROPHETESS OF ISRAEL: THE 22 COMMANDMENTS

PINUP

SEXIEST CREATIVE GENIUS IN HUMAN HISTORY

Artists March to the Beat of a Different Drummer
Sharon Esther Lampert Marches to the Beat of an Entire Orchestra

Poet, Philosopher, Prophet, Peacemaker
Paladin of Education, Princess & Pea
Phoenix, PHOTON, PINUP, Prodigy

Blue-Eyed. **Br**illiant. **B**eautiful. **B**uxom. **B**ooks

Sharon Esther Lampert was born an **OLD SOUL**—She was never young! Sharon is a lefty.

At age nine, her mother declared: "My daughter is a **p**oet, **p**hilosopher, and teacher!" She nicknamed her daughter, "The **P**rincess and **P**ea!"

Sharon's greatest literary works woke her up in the middle of the night—and made her get up out of bed — and write them down. Sharon writes an entire book in one day!

Sharon's mother was the sole person in Sharon's life who knew who she was from the **INSIDE OUT!** Her beloved mother also knew to her very last breath... the exact day and to the minute when she would die! (Eve Paikoff Lampert: June 3, 1925—May 5, 1985).

Later in life, Sharon will purchase a **green-pea pendant**, at the Broadway show, **"Once Upon a Mattress"** starring Sarah Jessica Parker. She wore the **green pea** every day around her neck with a beautiful Jewish-star pendant purchased in Haifa, Israel.

Sharon Esther's Gifts Are Metaphysical—Beyond the Scope of Scientific Inquiry

There Are No Rough Drafts!—Revelation: The Books Write Themselves!

Sharon Has 4 Books with GOD in the Title

"A LIST" Sharon Esther Lampert is One of the World's Greatest Poets
http://famouspoetsandpoems.com/poets.html

#1 Poetry Website for Student Poetry Projects

On a global scale, Sharon's poetry is used by teachers for their poetry lesson plans, and by students for their poetry school projects (see student **FAN MAIL**).

New York University Awards—YouTube Videos

Sharon Esther earned three degrees from **NYU**—and she was honored with two NYU awards. Sharon represented her class at her graduation—and was honored with an award for **"Multi-Interdisciplinary Studies."** She also played on the **NYU** Women's Varsity Basketball team as a center in the $16-million Coles Sports Center. Sharon won an **"NYU** Weightlifting Contest"—Sharon was the sole contestant—so she won! (**Washington Square News** article).

PINUP

Sharon Esther Lampert

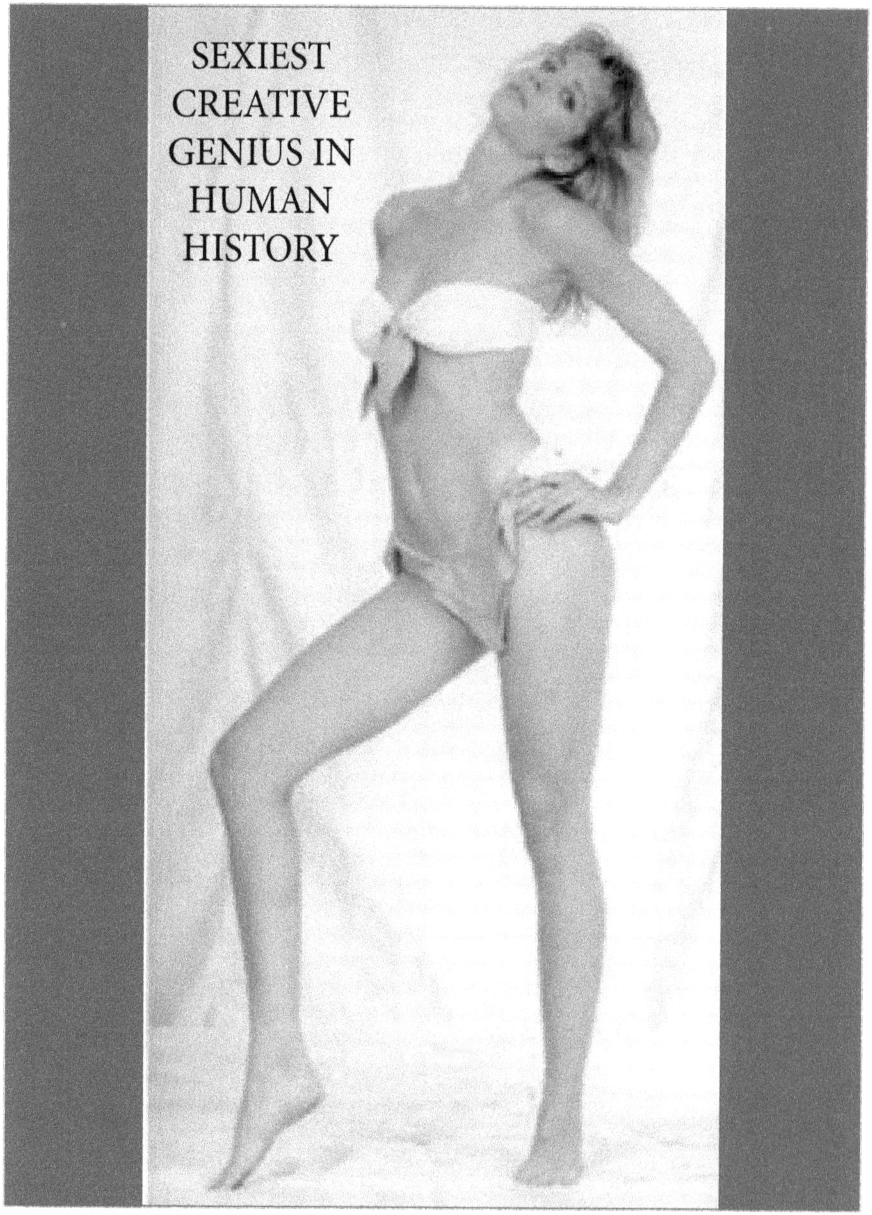

MYLIFE Is an OPENBOOK, to KNOWME Is to README

1.
PRODIGY
10 Esoteric Laws of
Genius & Creativity

Awesome Art of
Alliteration Using
One Letter of The
Alphabet

2.
PROPHET
22 COMMANDMENTS
A Universal Moral
Compass For All People

Who Knew **GOD** Was
Such a Chatterbox
A WORKING DEFINITION OF GOD
GOD IS GO! DO!

3.
POET
The Greatest Poems
Ever Written on
Extraordinary
World Events

POETRY WORLD RECORD

4.
PHYSICIST
LAWS OF INEXTRICABILIT

5.
PSYCHOBIOLOGIST
SPERM MANIFESTO

6.
PHILOSOPHER QUEEN
WOMEN HAVE ALL THE POW
PHILOSOPHY OF LOVE
PHILOSOPHY OF EVIL

7.
PHOTON SUPERHERO &
PALADIN OF EDUCATION
SMARTGRADES
BRAIN POWER REVOLUTIO

FAN MAIL

FANS@SharonEstherLampert.com

A PHENOMENON...
SHARON ESTHER LAMPERT

Lithe and lovely ... like a fawn.
This lady fascinates me ... from dusk till dawn.
Feminine and comely ... she's beyond belief
A blue-beam from her eyes ... is my soothing relief.

Girlish in her braces ... maidenly in her style
I yearn for her embraces ... and adore her friendly smile.
As tasteful as any artist ... you'll ever see
She's a compendium of class ... from A to Z.

If you'd like to see a fiqure, that puts Venus to shame
Behold her in a swimsuit, and your passions will aflame.
Ever exuding goodness . . . guided from above
Miss Sharon is the essence, and epitome of Love.

She's the inspiration of sages, and also fools like me
And the most magnificent female, I'm sure I'll ever see.
The nights are now endearing, & never filled with doubt
I sometimes wake up singing, cause it's Sharon . . .
I dream about.

Affectionately, . .
A devoted fan,
—Harry McVeety

SOLVE ONE PROBLEM EDUCATION SAVE ENTIRE WORLD

Sharon Esther Lampert

PALADIN OF EDUCATION & PHOTON SUPERHERO

My book was chosen as
BOOK OF THE MONTH
by the Alma Public Library in Wisconsin:

"The Silent Crisis Destroying America's Brightest Minds."
ISBN: 978-1885872548

My Pen Name:
Sharon Rose Sugar

SMARTGRADES
BRAIN POWER REVOLUTION

Smartgrades.com
EverydayanEasyA.com
BooksNotBombs.com

- 15 Stepping Stones of Academic Success
- 15 Stumbling Blocks of Academic Failure
- 10 Learning Tools for Academic Excellence
- 40 Universal Gold Standards of Education
- Parenting for Academic Success

FAN MAIL

FANS@SharonEstherLampert.com

Dear Sharon,

You are not only an exquisite poet, you're beautiful! Am smitten by your luminous beingness. Are you an angel in disguise--a so-called malachim in Hebrew if I am not mistaken.

Thank you for your wondeful open-hearted response. Your photo will sit next to those of Gautama Buddha and the Blessed Virgin Mary.

I will follow your sound esoteric advise regarding the positioning of your photo and the two other icons. I am deeply impressed that you are very conscious about the concept of sacred space and the flow of spiritual energy. So please send me your precious photo as soon as possible.

P.S. Will you be generous enough to send me your signed photo which I will place on the secret altar of my heart, lit by the menorah, the seven-stemmed candelabra of your inspiration, O mystical muse, O Rose of Sharon...

Your ardent fan and admirer,
—Felix Fojas, the cybercat with a mystical meow
Chico, CA, 95926

KADIMAH PRESS Gifts of Genius

Genius: Gifts of Divine Relevation: My Books Write Themselves!

Poet: The Greatest Poems Ever Written on Extraordinary World Events (25+Books)
Title: I Stole All the Words from the Dictionary
ISBN Hardcover: 978-1-885872-06-7
ISBN Paperback: 978-1-885872-07-4
ISBN E-Book: 978-1-885872-08-1

Prophet: **GOD IS GO! DO!**
Title: WHO KNEW GOD WAS SUCH A CHATTERBOX
A Working Definition of God
ISBN Hardcover: 978-1-885872-33-3
ISBN Paperback: 978-1-885872-34-0
ISBN E-Book: 978-1-885872-36-4

Prophet
Title: The 22 Commandments: All You Will Ever Need to Know About God
A Universal Moral Compass For All People, For All Religions, For All Time
ISBN Hardcover: 978-1-885872-03-6
ISBN Paperback: 978-1-885872-04-3
ISBN E-Book: 978-1-885872-05-0

Physicist
Title: NOW YOU KNOW EVERYTHING ABOUT EVERYTHING
ISBN Hardcover: 979-8-3492-9552-2
ISBN Paperback: 979-8-3492-9555-3
ISBN E-Book: 979-8-3492-9553-9

Prodigy
Title: Unleash the Creator The God Within: 10 Esoteric Laws of Genius and Creativity
ISBN Hardcover: 978-1-885872-21-0
ISBN Paperback: 978-1-885872-22-7
ISBN E-Book: 978-1-885872-23-4

Published 80+ Books

Prodigy: 6 Books
The Awesome Art of Alliteration Using One Letter of the Alphabet
Title: CUPID: Language of Love—Written in Letter C
ISBN Hardcover: 978-1-885872-55-5
ISBN Paperback: 978-1-885872-56-2
ISBN E-Book: 978-1-885872-57-9

Popular: Children's Book, Ages 8-12
Title: SCHMALTZY: IN AMERICA, EVEN A CAT CAN HAVE A DREAM
ISBN Hardcover: 978-1-885872-39-5
ISBN Paperback: 978-1-885872-38-8
ISBN E-Book: 978-1-885872-37-1
Schmaltzy.com

FIRST BOOK OF
COLOR-CODED
VOCABULARY WORDS

Popular
Title: SILLY LITTLE BOYS: 40 RULES OF MANHOOD
HOW DO SILLY LITTLE BOYS GROW INTO SANE BIG MEN
14 Global Catastrophes of Violence Against Women
ISBN Hardcover: 978-1-885872-29-6
ISBN Paperback: 978-1-885872-35-7
ISBN E-Book: 978-1-885872-41-8

Popular: EVERY RELATIONSHIP BEGINS WITH THE PERFECT MEAL
Title: SEX ON A PLATE: FOOD AS FOREPLAY
THE COOKBOOK OF EVERLASTING LOVE
ISBN Hardcover: 978-1-885872-46-3
ISBN Paperback: 978-1-885872-48-7
ISBN E-Book: 978-1-885872-47-0

THANK YOU

Count Your Blessings. Practice Gratitude.

Blessing 1. My Genetics, Two Sets of Artistic Genes and Gift of Genius: LEFTY—Like MOMMY!
- Genetic Inheritance: Painter Maternal Grandfather Benjamin Paikoff & Sculptor Father Abraham Lampert (Exhibit in Museum of Jewish Heritage, NYC)
- Vocalist: Ashira Orchestra, 18 Years: Ramaz Women's Service (YouTube video)
- Athlete: "Faster Than Any Boy, Anytime, Anywhere, Any Age!"

Blessing 2. My Life: Dawn of Digital Revolution
- APPLE: The Golden Age of Personal Computers
- ADOBE: The Golden Age of Creativity
- INGRAM: The Golden Age of Publishing
- SOCIAL MEDIA: The Golden Age of Internet & Global Communication
- iTUNES: The Golden Age of Music and Lyrics

Blessing 3. My Loved Ones
- **SELF-LOVE**: "Mindfulness, Meditation, Mantra, and Music Mitigates Madness!"
- Unconditional True Love: My MOMMY Eve Paikoff Lampert
- My PURRfect Children: SCHMALTZY and FALAFEL, Schmaltzy.com
- My "Friends First and Forever, and Family" NYU Tisch Professor Karl Bardosh
- My Metaphysical Sister: Poet on a Mission Hannah Sezenes, "ELI, ELI"
- My 7 Practice Husbands, Artist—Muses, Dates, and NYC Night Life
- My Bubbe Esther Tulkoff, EstherTulkoff.com

Blessing 4. My NYU Education and Awards
My Solomon Schecter Day School, Jewish Theological Seminary of America
My NYU Education: B.A., M.A., M.A. and AWARDS (YouTube video)
- NYU Professor Laurin Raiken
  NYU "Multi-Interdisciplinary Award" and M.A. Class Representative at Graduation
- Rockefeller University, NYC, Publication: "Hyperphagia and Obesity Induced by Neuropeptide Y" — Lab of Dr. Sarah Leibowitz and Dr. Glen Stanley
- AWARD: 100-Year Scholarship Award Winner, Presented by NYC Mayor Edward Koch
- AWARD: Empire Science Scholarship Award Winner
- AWARD: Jerusalem Fellowship Award, Aish Hatorah, Israel
- AWARD: First Prize: Upper East Side Resident Writing Contest
- Voice Teachers: Andy Anselmo of The Singer's Forum, NYC & Estelle Leibling
- Cantor Sherwin Goffin of LSS & Riva Alper of RAMAZ Women's Service (18 Years)

Blessing 5. My Sports
- NYC Marathon
- Basketball: NYU Women's Varsity Basketball Team, Center, NYU Coach Sherri Pickard
- NYU Weightlifting Contest Winner! $16 Million Coles Sports Center (solo contestant—so I won!)
- Basketball: NYC Urban Professional League
- B-Ball Coaches Chicago Bulls Phil Jackson and Boston Celtics Bill Walton, Omega Institute, NY
- Basketball and Softball—Coach Sandy Pyonin
- Skiing: Heavenly, Lake Tahoe, Nevada
- Tennis: NYC Central Park Tennis Courts
- Baseball, Hall of Fame: Coaches Wilma Briggs and Jean Harding, Omega Institute, NY

Blessing 6. My Inspirations
- ISRAEL: **"AM YISRAEL CHAI!"**
  Lambs to Slaughter to Lions & Light of the World: 22% of Nobel Prizes!
- NYC: The Golden Age of Personal Freedom & Creative Self-Expression
- AMERICA: Land of Unlimited Possibility, Potential, and Prosperity!

NYU Gallatin Professor Laurin Raiken and Me at My NYU M.A. Graduation

NYU Tisch Professor Karl Bardosh and Me Friends First & Family Forever!

LITERATURE IS POWERFUL BEYOND WORDS FOR IT CREATES WORLDS

— Sharon Esther Lampert

EVERY THOUGHT IN YOUR HEAD WAS PUT THERE BY A WRITER

— Sharon Esther Lampert

Please Keep in Touch!
Website, Facebook, Twitter, Instagram, YOUTUBE, Pinterest

FANS@SharonEstherLampert.com

Published 80+ Books

GENIUS: GIFTS OF DIVINE REVELATION

MY BOOKS WRITE THEMSELVES

I Am Mortal
MY BOOKS ARE IMMORTAL
Please Handle My Books Gently
My Books Are My Remains

This book was compiled in four parts:
Part 1. Birth of Idea: 2019
Part 2. Format Book: January 2022 & 2024
Part 3. Essays: October 2024
Part 4. Publish: November 2024

Sharon Esther Lampert

SEE THE WORLD THROUGH THE EYES OF A CREATIVE GENIUS
Prodigy, Prophet, Philosopher, Paladin of Education, Poet, Peacemaker, PINUP

FANS@SharonEstherLampert.com

www.ingramcontent.com/pod-product-compliance
Lightning Source LLC
LaVergne TN
LVHW071334210926
848486LV00003B/144